Soils, Rocks, and Landforms

Developed at
The Lawrence Hall of Science,
University of California, Berkeley
Published and distributed by
Delta Education,
a member of the School Specialty Family

1325245
978-1-60902-038-5
Printing 2 — 6/2012
Quad/Graphics, Versailles, KY

Table of Contents

What Is Soil?

Have you ever dug a hole in the ground? What did you remove to make the hole? **Soil**. Sometimes people call it dirt, but a scientist calls the layer of diggable material that covers planet Earth soil.

What is soil? If you pick up a handful of soil and look at it closely you might be able to see and feel what soil is made of. Soil is mostly made of several sizes of **rock**. You might see pebbles and smaller pieces of gravel. Soil usually contains sand. **Particles** of sand are really tiny rocks. Some pieces of rock are even smaller than grains of sand. Smaller pieces are called **silt**. The smallest pieces of rock are clay particles. Clay particles are too small to see, but you can feel them. Clay feels slippery when it is wet.

So soil is a mixture of different-sized rocks (pebbles, gravel, particles of sand, and even smaller particles of silt and clay) along with water and air. Rocks, air, and water are **earth materials**. But there is more to soil than earth materials.

Digging into soil

Soil is mostly made of rock in several sizes along with water and air.

Rock Size Chart	
Particle name	**Average size**
pebble	⬭
gravel	⬭
sand	∘
silt	·
clay	invisible to bare eye

3

Soil also contains organic material. Organic material is the remains of dead plants and animals. Plants send their roots into the soil and animals dig into the soil. When plants and animals die, their remains become part of the soil. Plants and animals **decay** into tiny pieces called **humus**. Humus provides **nutrients** for plants. Humus also helps the soil **retain** water.

What is an animal that lives in soil? Worms! Worms are good for soil and help plants grow. Worms burrow through the soil. As they move, worms mix the soil and make passageways for air and water. Worm waste also adds nutrients that are good for soil and plants.

A worm in soil enriched with humus

Not all soils are alike. Some kinds of soil have more humus. Some soils have more clay. Some have more sand, pebbles, and gravel.

Digging into Soils

1. What differences do you see in the soils shown above?

2. Where do you think these soils are found?

Weathering

Pebbles and sand are pieces of rock. Pebbles are pretty big. You can count a handful of pebbles. Pieces of sand are tiny. You can't count the particles in a handful of sand. All pebbles and sand particles start out as huge masses of rock the size of mountains. How do mountains break down into pebbles and sand?

The answer is **weathering**. Weathering is the breaking apart of rocks into smaller pieces. Weathering happens to all rocks when they are exposed to water and air.

Physical Weathering

Rocks break down in two ways. **Physical weathering** makes rocks smaller, but does not change the rocks in any other way. When a big rock falls from the side of a cliff, it breaks into lots of smaller rocks. All the **minerals** in the small rocks are the same as the minerals in the big rock.

When rocks get hot and then cold, they can crack. Sometimes water gets into cracks in rocks. Water expands when it freezes. It can expand enough to break big sections of rock along the crack. When ice melts, the rock may break into smaller pieces.

Physical weathering of cliffs

A rock weathered by freezing and thawing of water

Roots of trees and bushes can grow down into cracks in rocks. As roots grow, they make the cracks bigger. Sometimes the cracks get so big that the rock falls apart.

When rocks bang into one another, they get worn down. Rubbing, grinding, and banging is called **abrasion**. Abrasion is a kind of physical weathering. It happens when rocks fall in **landslides**, tumble in flowing water, or crash around in waves. Wind can blow sand against rocks. This sandblasting weathers the rocks.

Tree roots grow and break rocks.

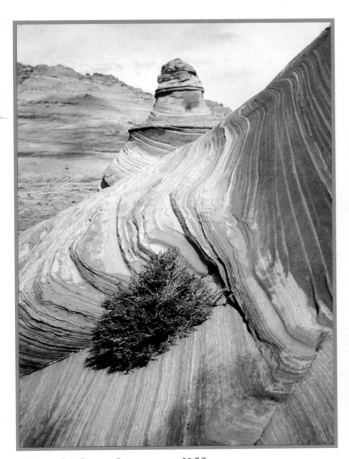

Sand abrasion on cliffs

Blowing sand can weather rocks into interesting shapes.

Chemical Weathering

Chemical weathering happens when minerals in rocks are changed by chemicals in water and air. The starting minerals change into new substances.

Many rocks contain iron. When oxygen in air comes in contact with iron, the iron in the rock can rust. Rust is iron oxide. Iron oxide is softer than other iron minerals. This causes the rock to break apart faster.

Carbon dioxide gas in the air **dissolves** in water droplets. This makes **acid**. The acid droplets can fall as rain. The acid causes the **calcite** in **limestone** and **marble** to make holes. This is a chemical change. Monuments, buildings, and gravestones made of marble or limestone change and weaken when exposed to acid rain.

Salt can cause chemical weathering. Salt water can **react** with minerals in rocks to make new minerals. When the new substances are softer than the original mineral, holes can form. The weak rock breaks and falls apart more easily.

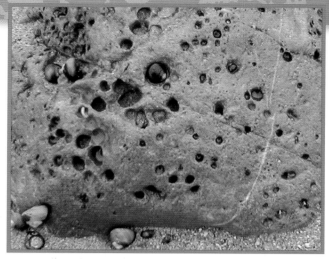

Chemical weathering of a rock containing iron

Chemical weathering of marble by acid rain

Chemical weathering of sandstone by salt water

Erosion and Deposition

A trip to the beach is fun. One of the best parts is playing in the sand. And there is so much sand. Where did it all come from? Was it made right there, or did it come from some other place?

Much of the sand on the beach came from mountains. **Erosion** moved the sand from the mountains to the beach. Erosion is the taking away of weathered rock. After rocks have weathered into small pieces, they can be carried away by gravity, water, or wind. Most of the sand shown here was carried to the beach by water flowing in rivers and streams.

As long as water keeps flowing, the bits of sand keep moving downstream. When the river enters the ocean, the water slows down. The sand settles to the bottom of the ocean. The settling of **sediments** is called **deposition**. Deposits of sand form beaches all over the world.

A beach

Erosion

The beach sand might start on high mountain cliffs. Sometimes big chunks of rock fall off the sides of mountains. Gravity pulls rocks downhill. Other times landslides move rocks and soil downhill.

Rainwater moving over the ground erodes the broken rocks. Water **transports** rocks into creeks. Water flowing in creeks transports broken rocks downstream. This process is called erosion.

Cliffs high in the mountains   **Weathered rock in a mountain creek**

Strong river currents move rocks downstream.

Creeks flow into rivers. Rivers have strong currents. Rivers can carry many sizes of rocks. The rocks bang together and rub on the riverbed. The rocks break into smaller and smaller pieces. The smaller pieces are pebbles, gravel, sand, and silt. Erosion continues. The farther the rocks move in the river, the smaller they get. They also get smoother and rounder as they tumble along.

Smooth, round pebbles along a river

Deposition

When the water flowing in a river slows down, the rocks are deposited as sediments. Large rocks are the first to settle to the bottom. Powerful **flood** waters move rocks of all sizes, even large boulders.

Where a river flows into a lake, a bay, or the ocean, the water slows down. Sand is deposited near the mouth of the river. The sand can form sandbars, deltas, and beaches. Farther out are deposits of silt and clay.

Large and small sediments deposited after a flood

Can you see deposits of sand and silt where this river enters the lake?

Can you see meanders in the river?

Other Kinds of Erosion and Deposition

Wind blows sand and smaller pieces of rock from one place to another. Sometimes the wind blows hard enough to carry a lot of sand and dust. Wind can erode valuable farmland.

When the wind dies down, sand and dust are deposited far from their starting places. This is how sand dunes form. Death Valley in California and Great Sand Dunes in Colorado are two places where large sand dunes formed.

Strong winds move earth materials from one place to another.

Sand dunes in Death Valley National Park, California

Great Sand Dunes National Park, Colorado

A U-shaped
valley eroded
by glaciers

Glaciers are frozen rivers. Rocks can be frozen in glaciers high in mountain canyons. Glaciers flow slowly through canyons. The frozen rocks scrape the floor and sides of the canyon. Glaciers weather and erode V-shaped canyons into U-shaped valleys.

Thousands of years ago in the Western United States, glaciers scraped down mountain valleys. They crushed and ground up rock beneath them. At the same time, glaciers covered much of the Midwest. These sheets of ice were over 1.5 kilometers (km) thick. They changed much of the landscape by eroding the surface and depositing the rock material in new places.

What happens when sand finally makes it to the ocean? Is that the end of the erosion and deposition story? Not quite. Waves erode beaches and deposit sand in different places all the time. As waves crash on the beach, sand continues to weather. Sand gets finer and finer. Sand abrades the rocks and cliffs along the ocean shore. Erosion and deposition go on and on.

**Sand deposited on a beach
around a weathered rock**

Reviewing Erosion and Deposition

1. Describe and give examples of erosion.
2. Describe and give examples of deposition.

Landforms Photo Album

Landforms Formed by Weathering and Erosion

Arch A curved rock that forms when chemical and/or physical weathering weakens the center, and the rock erodes.

Arches can form on the land or near the coast where waves batter and erode the center of the rock.

Butte A hill with steep sides and a small, flat top. A butte is smaller than a mesa.

Mesa A single, wide, flat-topped hill having at least one steep side.

Gorge A narrow, steep-sided valley or canyon.

Valley A low area between mountains where a stream or glacier flows. Stream valleys are V-shaped. Glacier valleys are often U-shaped.

Hanging valley A valley floor above another valley floor. Glacial erosion causes hanging valleys.

Canyon A V-shaped gorge with steep sides eroded by a stream.

Meander A curve or loop in a river or stream.

Hoodoo A rock shaped like a mushroom or statue. Hoodoos are formed when weak rocks erode away and leave behind stronger rocks.

Exfoliation dome A dome formed when rocks like granite peel away at Earth's surface.

Spheroidal rocks Rounded rocks formed by physical and chemical weathering.

Landforms Formed by Deposition

Alluvial fan A fan-shaped deposit of rocks formed where a stream flows from a steep slope onto flatter land.

Beach An area made of sand and other sizes of rocks between the low-tide and high-tide levels at the coast or a lake.

Floodplain Land covered by water during a flood. Small particles, like sand and silt, are deposited on a floodplain.

Delta A fan-shaped deposit of earth materials at the mouth of a stream.

Sandbar A long ridge of sand in shallow water, built up by river currents or ocean waves.

Levee A bank along a stream that may stop land from flooding. Levees can be natural or made by people.

Moraine The unsorted rocks and soil carried and deposited by a glacier.

Outwash plain A flat or gently sloping surface made of sorted sediments deposited by water from melting glaciers.

Plain A low area of Earth's surface that is often formed by flat-lying sediments.

Sand dune The sand deposited by wind in ridges, mounds, or hills.

Landslide The rapid downslope movement of earth material.

Slump A downward movement of a single mass of earth material.

Landforms Formed by Eruptions

Volcano A place where lava, cinders, ash, and gases pour out through openings in Earth's surface.

Caldera A hole that forms when the top of a volcano blows off or when the magma below the volcano drains away.

Cinder cone A volcano formed from a pile of cinders and other volcanic material blown out in an explosive eruption.

Composite volcano A volcano built by alternating eruptions of lava, cinders, and ash. Mount Rainier and the other volcanoes in the state of Washington are composite volcanoes.

Shield volcano A volcano built of very fluid lava. It looks wider than it is tall. Shield volcanoes created the Hawaiian Islands.

Landforms Formed by Crust Movements

Fault A break in Earth's crust where blocks of rock fracture and move. The San Andreas Fault has created a wide crack in Earth's surface.

Plateau A high, nearly level, uplifted area composed of horizontal layers of rock. The Colorado River has eroded the Colorado Plateau, forming the Grand Canyon.

Mountain A high, steeply sloped area where rock is uplifted along a fault or created by a volcano.

It Happened So Fast!

Some **landforms** are so old, you might think they've always been that way. The Sierra Nevada has been uplifting for more than 2 million years, and it continues today. The Colorado River continues to carve the Grand Canyon, as it has for more than 5 million years. Rock in the Appalachian Mountains began folding more than 480 million years ago. Most changes to Earth's surface are so slow, we can't see them happen.

But sometimes changes happen rapidly. Rapid changes affect people and landforms. Here are some examples of fast changes to Earth's surface.

Yosemite Floods of 1997

Floods caused a lot of damage in northern California in 1997. Three factors created the floods. There was deep snow in the mountains, warm temperatures, and heavy rain.

Yosemite National Park was hit hard. The Merced River, which flows through Yosemite Valley, rose higher than ever before. Water spread out and covered much of the valley.

Flood water in Yosemite

A sign showing the water level during the flood

A Yosemite hiking trail after the flood

In some places, the flood water was 3 meters (m) deep. Campsites were washed away. Housing for the people who worked in Yosemite was destroyed. When the water slowed, sand and other sediment were deposited all over the valley floor. The course of the river was different. Flood water has the power to change the land rapidly.

Water flowing out of Canyon Lake over the spillway at the height of the flood

Canyon Lake Flood of 2002

In the 1960s, a dam was built on the Guadalupe River to prevent flash flooding and to provide a water supply for central Texas. The Canyon Dam created a reservoir called Canyon Lake. The lake became a popular recreation area in the Texas Hill Country between San Antonio and Austin. This area of Texas is known worldwide for its potential for flash floods. In fact, it is part of an area called Flash Flood Alley.

In early July 2002, it began to rain heavily in the upper part of the Guadalupe River watershed. About 1 m of rain fell during 1 week, and it continued to rain. The runoff flowed into the river and down to Canyon Lake. The lake was already full because of the rain falling on it. The water began to flow over the spillway of the dam. A wall of water over 2 m high and 380 m wide went over the dam. For the next 6 weeks, an amount of water equal to one and a half times the amount of water in the lake flowed over the spillway and into a narrow valley behind Canyon Dam. This torrent of water cut into the valley floor. It washed away the oak trees, mesquite, and topsoil and carried tons of sediment downstream.

When the water stopped flowing, the valley became a gorge. The gorge is 1.6 kilometers (km) long, hundreds of meters wide, and up to 15 m or more deep. The new walls of the gorge exposed limestone rocks dating back 100 million years. Those rocks contain **fossils** of worms and crustaceans, and tracks of ancient insects and dinosaurs. Teams of scientists are carefully observing the area and recording the evidence they find. The limestone is very brittle and breaks away from the canyon walls.

The other feature that was exposed by the flood was an **earthquake** fault. The **fault** was known to be there before the flood, and now 800 m of the Hidden Valley Fault is visible. The gorge has become a fresh, new laboratory for **geologists** to study faults found in limestone.

The Gorge Preservation Society (GPS) is a local citizen's group that works with the Guadalupe–Blanco River Authority (GBRA) and the U.S. Army Corps of Engineers to protect and study the gorge. They lead public hikes to some parts of the gorge.

The Canyon Lake flood formed this gorge.

Big Thompson Canyon Flood of 1976

The state of Colorado was celebrating its centennial on July 31, 1976. People were enjoying their summer vacations camping along the Big Thompson Canyon. The canyon is northwest of Denver. The town of Estes Park is at the western end of the canyon.

Thunderstorms often occur in the Rocky Mountains, especially in the afternoon. On this day, a thunderstorm formed over the western end of Big Thompson Canyon and didn't move. It dumped over 30 centimeters (cm) of rain in less than 4 hours. The canyon is steep and narrow, and there is little soil to retain the water. By 9:00 p.m. that evening, a wall of water more than 6 m high roared down the canyon. It was a flash flood! It sped down the canyon at about 6 m per second. Huge boulders were swept down the canyon by the wall of water.

People in the lower parts of the canyon had no warning. Because the flood happened so fast, the only way for people to escape was to climb to higher ground in the canyon. Many people didn't have time to get out of the way of the water. There were 145 deaths. A lot of things were destroyed, including 400 cars, 418 houses, and 52 businesses. Most of the main highway along the canyon was washed out. More than 800 people were evacuated by helicopter the next morning. The Big Thompson Canyon Flood was one of the deadliest flash floods ever reported in the United States.

Since this event in 1976, early-warning systems have been put in place. This advance notice about possible flash floods helps people move to safety.

Many houses were destroyed in the Big Thompson Canyon Flood.

Mount St. Helens Eruption of 1980

Mount St. Helens is a **volcano** in the state of Washington. It is part of a chain of volcanoes called the Cascade Range. On March 20, 1980, an earthquake happened. The north side of the mountain started to bulge.

Two months later, with little warning, another earthquake happened. On May 18, at 8:32 a.m., a magnitude 5.1 earthquake shook Mount St. Helens.

The bulge disappeared as a large avalanche of rocky debris slid down the side of the volcano. **Pumice** and ash erupted.

The debris filled the valley below Mount St. Helens. Trees toppled over like toothpicks. Over 600 square km were flattened. Volcanic mudflows called lahars spilled into the rivers. There were 57 people killed. An estimated 12 million fish at a hatchery, and 7,000 deer, elk, and bears were also killed. The eruption destroyed or damaged over 200 homes, 27 bridges, 298 km of highway, and 24 km of railways.

Mount St. Helens after the 1980 eruption

The May 18 eruption lasted more than 9 hours. The plume of ash reached to over 20 km above sea level. It moved eastward at an average speed of 100 km per hour. Ash traveled as far as Idaho by noon the next day. It was also found on tops of cars and roofs in Edmonton, Alberta, Canada, the next morning.

Since 1980, Mount St. Helens has had several smaller eruptions. It appears that the volcano is not yet ready to be dormant.

A view of the ash in Connell, Washington, after the eruption. Ash fell over 57,000 square km.

Two scientists standing by fallen trees in Smith Creek valley

Homes and highways damaged in the Northridge Earthquake

Northridge Earthquake of 1994

At 4:30 a.m., on January 17, 1994, people living in the Los Angeles, California, area got a jolt. It was an earthquake deep under the city of Northridge, California. The earth shook for 15 seconds. The magnitude of the earthquake was 6.7.

Earth's **crust** has a lot of cracks. The cracks are called faults. Earthquakes happen when huge sections of Earth's crust slide past each other on a fault.

The Northridge Earthquake happened on a fault geologists didn't know about. It was a blind thrust fault. Blind thrust faults don't reach all the way up to Earth's surface. They are hidden faults.

Damage was widespread. Sections of major highways fell. Parking structures and office buildings fell apart. Many apartment buildings were beyond repair. Houses in the towns of San Fernando and Santa Monica were also damaged. About 22,000 people lost their homes.

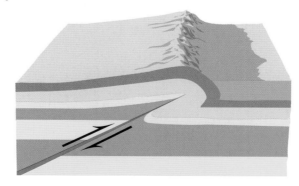

A diagram of a blind thrust fault

San Francisco soon after the 1906 earthquake, with smoke rising in the background

San Francisco Earthquake of 1906

Wednesday, April 18, 1906, was the day a big earthquake struck San Francisco, California. People felt the first little shakes at 5:12 a.m. Soon after, the major shaking started. It lasted 47 seconds. People as far away as southern California, Oregon, and central Nevada felt it. The magnitude of the earthquake was about 7.9.

Movement on the San Andreas Fault caused the 1906 earthquake. The fault broke at Earth's surface for a distance of 470 km. Cracks opened and cliffs formed where sections of land fell. Land near the San Francisco Bay settled as a result of the shaking. The settling and shaking caused buildings to fall. Great fires broke out because gas pipes broke. And the fires burned because there was no water to put them out due to water pipe damage.

Earthquakes usually last only a few seconds. Earthquakes cause the fastest changes to landforms of any common natural event.

The town of La Conchita at the base of an unstable hill

La Conchita Landslides of 1995 and 2005

Landslides occur when rocks and soil quickly slide downhill. Some areas are more likely to have landslides. The hillside above La Conchita, California, is one of those areas. This small town has had two large landslides. The slides killed people and damaged buildings and cars.

The landslide shown here happened on March 4, 1995. Many people were evacuated because of the slide. Houses nearest the landslide were completely destroyed. No one was killed or injured.

People continued to live in the area below the hillside. Another landslide happened on January 10, 2005. It destroyed or damaged 36 houses and killed 10 people. This kind of land movement happens so quickly, it is often impossible to get out of the way.

This landslide is another example of erosion and deposition. You can see where the sand and mud eroded from the top of the hill. You can also see where the sediment was deposited on the edge of town.

Yosemite Rockfall of 1996

On July 10, 1996, Ernie Milan was jogging on a trail in Yosemite National Park. Ernie was a trail worker for the National Park Service, so he knew the area well. He heard a loud boom. Dust started swirling around him. Day turned into night. What happened?

A giant mass of **granite**, weighing nearly 70,000 tons, broke loose from a cliff. It fell 600 m to the valley floor. Hundreds of trees were knocked over. One person was killed, and several others were injured. Ernie was not hurt.

Scientists estimated that the rock hit the floor at 400 km per hour. There is no way to stop rockfalls or predict when they will happen again.

Rockfalls happen where loose or cracked rocks are on steep slopes. Rockfalls may happen along road cuts and other excavations. Rockfalls start when rocks are dislodged by freezing or thawing of water or by heavy rainfall. Rockfalls can also be triggered by ground shaking from earthquakes. They generally occur without warning.

Rockfalls happen all over the world. They are a natural kind of weathering and erosion. Most rockfalls are not observed by people. But scientists try to learn what they can from these rockfalls. Some day scientists might be able to predict when a mass of rock is ready to break away.

A 70,000-ton mass of granite fell to the floor of Yosemite Valley.

33

Where Do Rocks Come From?

Where do rocks come from? This question keeps geologists busy. Even though they don't have all the answers, they know a lot about where rocks come from.

Earth is about 4.6 billion years old. The oldest dated rock found on Earth is about 4 billion years old. That's almost as old as Earth. Scientists have also found crystals of a mineral called zircon that were formed 4.4 billion years ago.

There are three big groups of rocks: **igneous**, **sedimentary**, and **metamorphic**. All the rocks in a group have similar origins, often inside Earth.

Earth is like an egg. An egg has a hard outer layer called the shell. Earth has a hard outer layer called the crust. Earth's crust is made of solid rock.

Under an egg's shell is the fluid egg white. Under Earth's solid crust is the **mantle**, partly melted rock that flows like really thick toothpaste. It is hot inside Earth. It is so hot that rocks and minerals melt.

An egg has a yolk in the center. Earth has a metal **core** in its center. Earth's core is made of iron and nickel.

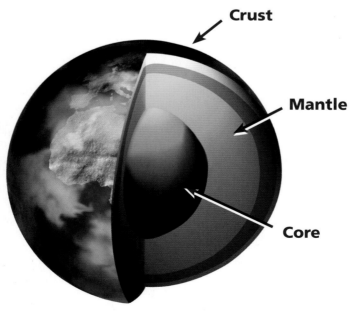

Crust

Mantle

Core

A cross section of Earth

Igneous Rocks

Igneous rocks start out as melted rock deep in Earth's crust. Sometimes the melted rock, called **magma**, comes to the surface in volcanoes. The magma pours out as **lava**. When lava cools and hardens, it forms new rocks. The basalt you tested for calcite is volcanic igneous rock. Much of the rock in the Cascade Mountains of the Pacific Northwest is basalt.

Other times, magma cools slowly and hardens below the surface. Earthquakes and other changes in Earth's crust might bring these igneous rocks to the surface years later. The granite you studied cooled below Earth's surface. The Sierra Nevada in central California and the Rocky Mountains in Colorado, Wyoming, and Montana are mostly granite mountains.

Basalt

Granite

Sandstone

Sedimentary Rocks

Sedimentary rocks form from bits and pieces of recycled rocks and minerals. **Sandstone** is an example of a sedimentary rock. Sandstone starts as big rocks in the mountains. Over time, the rocks crack and break into smaller pieces. This process is called weathering.

Water can cause weathering. Water freezes in cracks in rocks. It expands when it freezes and breaks the rocks apart. Tree roots also cause weathering. Roots grow into cracks in rocks and break big pieces of rock loose.

Loose rocks tumble downhill and break into smaller pieces. Pieces might end up in streams and rivers. The pieces get banged around and broken into smaller and smaller pieces. Eventually the rocks from the mountain are reduced to tiny pieces of sand.

Sand often gets deposited in the ocean and bays. Layers of sand build up. The layers of sand are called sediment. As millions of years pass, the sand gets buried under more layers of sediment. Sand particles are pressed and stuck together. The sand turns into the sedimentary rock sandstone.

Sedimentary rocks often have bits of sand and gravel you can see. Sometimes sedimentary rocks contain fossils of shells, animals, or plants. Sedimentary rocks form in layers. If the rocks are still in their natural site, you can often see the layers.

Sandstone layers

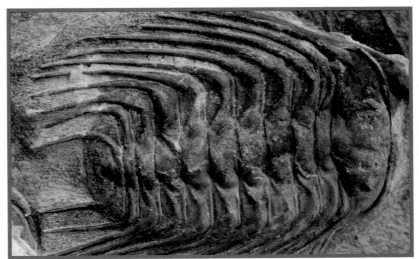

A trilobite fossil

A fern fossil

A shell fossil

Metamorphic Rocks

Meta- means change. *Morph* means shape or form. Metamorphic rocks change from one kind of rock into another kind of rock. The starting rocks can be igneous, sedimentary, or even other metamorphic rocks. The rocks change because of heat and pressure. If a rock gets buried deep in Earth's crust or touches hot lava, it will change into metamorphic rock.

Heat and pressure can turn sandstone into quartzite. Limestone can become marble. Shale can change into slate. Heat and pressure can turn granite into gneiss (pronounced "nice").

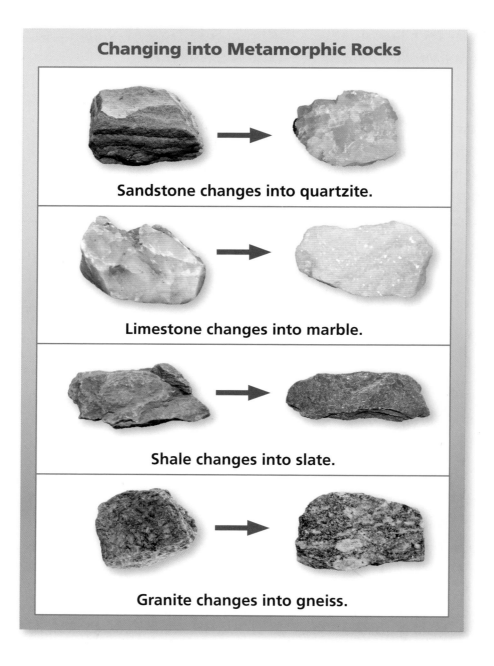

Changing into Metamorphic Rocks

Sandstone changes into quartzite.

Limestone changes into marble.

Shale changes into slate.

Granite changes into gneiss.

The Rock Cycle

Metamorphic rocks aren't the only rocks that can change. Over time, any kind of rock can change into any other kind of rock. The changes from igneous to sedimentary to metamorphic and back to igneous are the **rock cycle**.

For example, a piece of igneous granite might weather into sediments. The sediments can end up in a layer with other sediments. After a long time, the sediments might change into sedimentary sandstone.

The sandstone could get heated by a lava flow or buried under other sediments. The heat and pressure might change the sandstone into metamorphic quartzite. And finally, the quartzite might be carried down into Earth's mantle where it will melt. After millions of years, the rock material might come back in a new piece of igneous granite.

Rocks don't all follow this path through the rock cycle. The important thing to remember is that all rocks change. Any rock can change into any other kind of rock. Study the rock cycle illustration to see how.

It is even possible for a rock to re-form as the same kind of rock. For example, sandstone might weather into sand. The sand could pile up in a bay. After millions of years, the sand might become new sandstone.

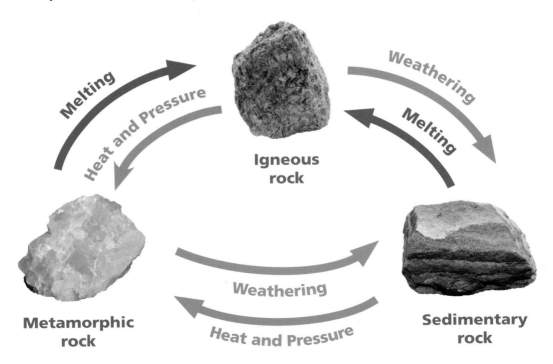

Any kind of rock can change into any other kind of rock. This is the rock cycle.

Rock Samples

This table shows several examples of sedimentary, metamorphic, and igneous rocks. How many of them have you held in your hand?

Sedimentary	Metamorphic	Igneous
Limestone	Marble	Basalt
Sandstone	Quartzite	Obsidian
Shale	Slate	Tuff
Conglomerate	Gneiss	Pumice
Breccia	Schist	Granite

Mohs' Scale and Birthstones

Geologists use a number scale to describe mineral **hardness**. Friedrich Mohs (1773–1839) came up with the scale. Mohs was a German scientist who studied minerals. He knew that some minerals could scratch others. If one mineral could scratch another, it must be harder than the mineral that was scratched.

Mohs' scale goes from 1 (softest) to 10 (hardest). The mineral talc, the softest mineral, has a hardness of 1. Diamond, the hardest mineral, has a hardness of 10. All the other minerals fall between talc and diamond. These ten minerals represent the ten levels of hardness.

Minerals with higher numbers on Mohs' scale will scratch minerals with lower numbers. Calcite can scratch gypsum, but calcite cannot scratch fluorite. **Quartz** can scratch **feldspar**, but quartz cannot scratch topaz.

Some minerals can be broken apart with a geologist's hammer.

Mohs' Scale

	10 Diamond
	9 Corundum
	8 Topaz
	7 Quartz
	6 Feldspar
	5 Apatite
	4 Fluorite
	3 Calcite
	2 Gypsum
	1 Talc

Birthstones

The minerals at the hard end of Mohs' scale are called **gems**. Gems are hard minerals that can be cut into beautiful shapes. Because they are hard, they last a very long time.

People have identified gem minerals with the months of the year. These gems are called **birthstones**. At one time, people believed birthstones could protect, heal, or bring good luck. Some birthstones were thought to have magical powers. Which of these beautiful minerals is your birthstone?

January: Garnet

Often colored red, garnet crystals form easily in rock. Long ago, people believed a garnet would protect its owner from wounds.

February: Amethyst

Amethyst is a form of quartz that is purple. Amethysts were once thought to keep soldiers safe in battle.

March: Aquamarine

The mineral beryl forms crystals of different colors. Blue crystals are called aquamarines. Aquamarines were supposed to bring good luck to sailors at sea.

April: Diamond

It takes extreme heat and pressure to turn carbon into diamond. Diamond is the hardest natural substance in the world. Some people believed a diamond would give them strength.

May: Emerald

Unlike other gems, emeralds are not found washed into streams. They must be mined from other rock. It was said that an emerald placed under the tongue would let you see the future.

June: Alexandrite

Along with pearl, alexandrite is the birthstone for June. This mineral is both rare and unique. In natural sunlight it looks green. But when it is lit by a lightbulb, it looks red.

July: Ruby

Most rubies are found in streambeds or under soil. It was thought that a ruby turning dark warned its owner of danger.

August: Peridot

Green and yellow-green peridots are forms of the mineral olivine. They can be found in lava flows and in meteorites. Peridot is the only gem that may come from outer space! Peridot was said to give its owner dignity.

Scratch Match

Here is a list of birthstones from hardest to softest based on Mohs' scale. Diamond is the hardest birthstone. What can you say about the hardness of the other birthstones from this data?

- Diamond (10)
- Ruby and sapphire (9)
- Alexandrite (8.5)
- Topaz (8)
- Aquamarine and emerald (7.5–8)
- Garnet (7–7.5)
- Amethyst and peridot (6.5–7)
- Opal (5.5–6.5)
- Turquoise (5–6)

September: Sapphire

Both sapphires (blue) and rubies (red) are crystals of the same mineral, corundum. Legend says a sapphire will make a foolish person wise.

October: Opal

Opals can be clear, cloudy, or **opaque**. Opals of good quality may show rainbowlike colors. The Romans wore opals for love and hope.

November: Topaz

Pure topaz is colorless. Impurities can make topaz yellow, blue, green, orange, or pink. In the Middle Ages, topaz was said to improve the mind. The largest known topaz crystal weighs 271 kilograms (kg)!

December: Turquoise

Bluish-green turquoise is an opaque mineral that is rarely found in crystal form. The Navajo people believed a turquoise thrown into a river with a prayer would bring rain.

Common Minerals

Quartz	Fluorite	Calcite	Gypsum
Cannot be scratched with steel	Can be scratched with steel	Can be scratched with aluminum	Can be scratched with fingernail
Hardness = 7	Hardness = 4	Hardness = 3	Hardness = 2
A common mineral in Earth's crust	Appears in many colors	A common mineral in Earth's crust	Used to make plaster

Identifying Minerals

Rocks are made out of earth materials called minerals. Minerals can be colorful and beautiful. Minerals come in many colors. Look at the photo above. The minerals are (1) copper, (2) malachite, (3) fluorite, (4) sapphire, (5) ruby, (6) sulfur, (7) elbaite, (8) silver, (9) beryl, and (10) galena.

Many kinds of rocks can be made out of just a few kinds of minerals. Geologists try to find out what kinds of minerals are in rocks. They use **property** tables to help them identify the minerals they find in rocks. Here are some of the mineral properties they use.

Fluorite is a mineral that comes in many colors.

Color

Some minerals, such as black biotite mica, have colors that help identify them. But many minerals come in several colors. The property of color does not always help identify a mineral.

Hardness

Hardness is determined by trying to scratch a mineral with different materials. Minerals with large hardness numbers, like 8 and 9 on Mohs' scale, can't be scratched except with a diamond. Even a steel knife won't scratch a hard mineral. Minerals with small hardness numbers, like 1 and 2 on Mohs' scale, are easy to scratch. A fingernail can scratch these minerals.

A scientist might use a nail to test the hardness of a mineral like this feldspar.

The mineral hematite leaves a reddish brown streak.

Streak

If you rub a mineral on a tile, it can leave some mineral powder. This is called **streak**. The color of the streak can help a geologist identify the mineral. Streak is usually a better property to use than the color of the mineral sample. Color varies within some minerals. But streak color is usually the same for all samples of one kind of mineral.

Luster

Luster describes the way light reflects off a mineral's surface. The simple way to describe a mineral's luster is **metallic** or **nonmetallic**. Metallic luster means the mineral shines like a metal. Nonmetallic means it does not shine like a metal. A nonmetallic mineral could be glassy, pearly, or dull. The pyrite shown below has a metallic luster. The fluorite and calcite have nonmetallic lusters.

Iron pyrite has metallic luster. Fluorite and calcite have nonmetallic lusters.

Cleavage surfaces are smooth and flat in this mineral.

Quartz fractures unevenly.

Obsidian has rounded fractures.

How Minerals Break

Regular, flat surfaces on broken minerals are called **cleavage**. Some minerals only **fracture**. That means when they break, the surfaces are uneven, rounded, or splintered. Calcite shows cleavage. Quartz and the rock obsidian show uneven and rounded fractures.

Magnetism

A mineral is **magnetic** if it is attracted to a magnet. Lodestone is a special form of the mineral magnetite that acts as a magnet itself. Lodestone was used in some of the first compasses because of its magnetic property.

Steel nails stick to this magnetic mineral.

Thinking about Mineral Properties

Use the Mineral Properties Table on page 49 to answer these questions.

1. How can you tell the difference between gold and pyrite?

2. How can you tell the difference between calcite and quartz?

3. How can you tell the difference between talc and gypsum?

4. A mineral sample is dark gray, can be scratched with a fingernail, has a metallic luster, and feels greasy. Which mineral is it?

Mineral Properties Table

Mineral	Color	Hardness	Streak	Luster	Break	Other Properties
Calcite	colorless, white, gray	3	white	nonmetallic	cleavage	Fizzes in acid Double refraction
Feldspar	white, pink, gray	6	white	nonmetallic	cleavage	
Fluorite	clear, purple, yellow, colorless, green	4	white	nonmetallic	cleavage	Glows under a black light
Galena	dark gray	2.5	gray	metallic	cleavage	Heavy for its size
Gold	yellow	2.5 to 3	golden yellow	metallic	fracture	Very heavy for its size
Graphite	dark gray to black	1 to 2	black	metallic	cleavage	Feels greasy
Gypsum	colorless, white, gray	2	white	nonmetallic	cleavage	Can form rosettes or fibers
Hematite	brown, red, black	5 to 6	reddish brown	metallic	fracture	Can appear glittery with a bright metallic luster, or dark brown or red with a dull luster
Hornblende	dark green, brown, black	5 to 6	white or gray	nonmetallic	cleavage	May have long crystals with parallel sides
Magnetite	black	5.5 to 6.5	black	metallic	fracture	Attracted to a magnet
Malachite	dark green, light green	3.5 to 4	pale green	nonmetallic	fracture	Silky luster Bubbles with strong acid
Mica	dark brown, black, white	2 to 3	colorless	nonmetallic	cleavage	Pulls apart in sheets
Pyrite	yellow	6 to 6.5	greenish black	metallic	fracture	
Quartz	colorless, white, rose, gray, purple, brown	7	colorless	nonmetallic	fracture	
Talc	white, greenish to gray	1	white	nonmetallic	cleavage	Feels greasy May pull apart in fibers

Mining for Minerals

Gold

Do you know what a forty-niner is? That's the nickname given to people who moved into California in 1849. Why did they go? **Gold**!

Gold ore— gold in quartz and a gold nugget

Gold was discovered in a stream in Coloma, California, in 1848. It was in the form of nuggets. People could scoop up a pan full of stream gravel and wash the gravel away. This would leave the heavier gold in the pan. In a few years the California gold country was full of thousands of gold panners trying to strike it rich. Before too long the gold was panned out.

Was all the gold gone? No. Nuggets were hard to find, but rocks still held a lot of gold **ore**. Ore is any rock or mineral that has a valuable substance in it. Gold ore has gold in it. Silver ore has silver. Iron ore has iron, and so on.

The search for gold changed from panning to mining. Gold miners used dynamite and shovels to get the ore. Sometimes they dug deep tunnels to get the ore. Big machines crushed the ore. The miners then separated the gold from the rest of the minerals.

Bauxite

Bauxite is a rock. It is made of three minerals. Each of the minerals has the metal aluminum in it.

In many places around the world, bauxite is a red clay. Bauxite deposits found on or near Earth's surface are scooped up with big power shovels. The ore is crushed, screened, and ground into powder. Then heat and electricity separate aluminum out of the bauxite.

Aluminum is three times lighter than steel and almost as strong. It is a good material for building bridges, airplanes, bicycles, and cars. Because it does not rust, aluminum is made into boats and siding for houses. It is also made into wire, cans, kitchenware, and foil for wrapping food.

Aluminum ore—bauxite

Iron

The most important metal in the world is iron. Modern buildings, train tracks, ships, tools, cars, and thousands of other things are made of iron. Well, not iron exactly. Most iron is made into steel. Steel is essential to modern industry.

Iron ore—hematite

Iron is made from iron ore. The most common iron ore is a mineral called hematite. Huge furnaces heat the ore with charcoal and limestone. This is called smelting. The iron melts. The liquid iron is poured into molds. The raw iron is called pig iron.

Pig iron is brittle, so it breaks easily. But adding carbon to the iron makes the iron strong and flexible. Iron with carbon is steel.

Steel can be rolled into sheets. Sheet steel is used to make car bodies, refrigerators, and many other things. Steel can be made into cutting tools with sharp edges. Steel can also be shaped into big beams to make large buildings. Nails and bolts that hold things together are made of steel.

A hematite mine

A furnace making pig iron

Copper

Copper is the metal that gives pennies their color. It is also the most important metal used to make electric wires. Copper is one of the best conductors of electricity.

Copper ore—malachite

Copper ores are minerals. One of the most common copper ores is the mineral malachite. It is bright bluish green. Where copper ore is found near Earth's surface, it is broken apart with dynamite. The pieces are then loaded into trucks and taken to the smelter.

At the smelter, the ore is put into a furnace. The heat melts the copper. It flows out the bottom of the furnace into molds. The copper is ready to be made into wire, cooking pots, and pipes.

When copper is mixed with another metal called tin, it makes bronze. Bronze has been used for centuries to make things like church bells and statues. Why?

Bronze bells make a beautiful sound when they are struck. But there is another reason. Bells and statues are outdoors all the time. They get wet. Bronze does not rust like iron. Objects made of bronze are not changed by weather. A bronze statue will last for thousands of years.

An open-pit copper mine

Loading copper ore

Monumental Rocks

Humans build monuments to honor important people and events. Monuments are built to last a long time. They are usually large structures. If you were going to build a monument, what would you make it out of? Rock would be a good choice.

From ancient times to the present, people have made monuments out of rock. Why? Because rock is found everywhere. Rock can be cut and shaped. And most of all, rock lasts a long time. Some rocks are nearly as old as Earth itself. Some structures made of rock have been standing for thousands of years.

The Great Pyramid

Did you know that the Great Pyramid in Egypt is almost completely solid? The only spaces inside are a few hallways and rooms. The pyramid is made out of about 2,300,000 blocks of limestone and granite! The average block weighs about as much as two cars. The largest blocks weigh as much as six cars.

As big as it is, the Great Pyramid was made to honor just one person. It was built around 2700 BCE to hold the body of the pharaoh Khufu. He was a ruler of ancient Egypt.

Each building block in the pyramid was pulled to the site on a wooden sled. Workers used copper axes, chisels, and saws to cut and fit the stones. Today, people wonder how such a building was made without iron tools.

Limestone for the pyramid's center was cut from nearby cliffs. That way, the stone did not have to be moved very far. Granite for the walls and doorways came from almost 1,000 kilometers (km) up the Nile River. Nicer-looking limestone for the outside of the pyramid came from a few kilometers away. By using barges, they floated the limestone down the river to the building site. Today, the polished limestone shell is gone. The stone was "recycled" in the 1300s to rebuild a city damaged by earthquakes.

What Is Limestone?

Limestone is a sedimentary rock that forms from calcium carbonate. Limestone forms under water. Tiny bits of calcium carbonate, some from shells of organisms, drift to the bottom of the ocean or bays. These pieces of calcium carbonate pile up for millions of years. The layer of calcium carbonate gets thicker and thicker. After a very long time, the bits of calcium carbonate turn into limestone.

The Taj Mahal

The Taj Mahal in Agra, India, is one of the wonders of the world. Many agree that it is the most beautiful building of all time. The Taj Mahal's designer was Shah Jahan (1592–1666). He built the monument to honor his wife, Mumtaz Mahal (1593–1631).

The Taj Mahal, which means "Crown Palace," is made entirely of white marble. Builders from all over the Middle East worked 22 years to make it. Inside the Taj Mahal, colorful marble was cut and pieced together like a puzzle. Forty-three different kinds of gemstones were used for decoration.

What Is Marble?

Marble is a metamorphic rock. Metamorphic rocks form when one kind of rock changes into another kind of rock. This usually happens when heat and pressure act on a rock for a long time.

Marble starts out as limestone. When limestone gets buried deep inside Earth, the pressure builds. The temperature goes up. After millions of years, the limestone changes into marble.

The Vietnam Veterans Memorial

The Taj Mahal and the Great Pyramid each honor a single person. The Vietnam Veterans Memorial was built to honor all the Americans who died in the Vietnam War (1955–1975). It was the idea of Vietnam veteran Jan Scruggs.

A competition was held to design the memorial. Maya Lin's (1959–) plan was chosen. At the time, Lin was 21 years old. She was a student at Yale University. Lin designed the monument as a black granite wall. The wall forms a V. She wanted the rock to rise out of the ground like two arms to embrace people. One arm points to the Washington Monument. The other arm points to the Lincoln Memorial.

The wall was finished in 1982. The names of more than 58,000 military men and women are written on the wall. Many people didn't like Lin's design at the time. But the memorial she designed is one of the most visited sites in Washington, DC.

Each year millions of people visit the Vietnam Veterans Memorial, designed by Maya Lin (right).

What Is Granite?

Granite is an igneous rock. That means it started as melted rock deep under Earth's surface. As the melted rock moved toward the surface, it cooled and crystallized. When you look closely at granite, you can see the crystals of the different minerals.

There are only a few places in the world where black granite is found. The beautiful black granite in the Vietnam Veterans Memorial comes from India.

The Washington Monument

George Washington died in 1799 at his home in Virginia. That year, Congress voted to move his body to the capital city. They wanted to bury it under a marble monument. But Washington's body was never moved, and 85 years passed before a monument was completed. During that time, politicians fought. There were problems raising money for the monument. Sometimes there weren't enough railway cars to deliver the marble. For 2 decades, work stopped completely.

After the Civil War (1861–1865), interest in the monument rose again. People were worried about it. They said the base was not strong enough to support the finished building. Some of the marble blocks were splitting. The original plans had been lost. Many thought the monument was ugly. Some wanted to knock it down and start over. Others wanted a new design. Finally work went on after the foundation was made stronger. The upper two-thirds of the structure was built.

The outside of the monument was constructed from marble taken from quarries in Maryland. At first, the two sections looked the same. But over time wind and rain have caused the marble sections to weather differently. Can you see a color difference between the original marble from one quarry and the later marble from a different quarry?

The great stone monument was finally completed in 1884. It is 169.3 meters (m) tall, which makes it the tallest structure in Washington, DC. It is also the tallest stone structure in the world!

Geoscientists at Work

Where do **geoscientists** go to do their work? The answer is, just about everywhere. That's because the prefix *geo–* means Earth. Geoscientists observe, investigate, and test locations all over Earth. Their job is to discover, manage, and protect Earth's natural materials. Earth's rocks, minerals, soils, water, air, plants, animals, and fossil fuels together are called **natural resources**. Geoscientists study the history, distribution, use, and conservation of Earth's limited, valuable, nonliving natural resources.

Most geoscientists spend a lot of time doing fieldwork. That means they are outside in direct contact with Earth. Most geoscientists have a specialty. They focus on one part of the Earth system.

Marine Geologists

Marine geologists study the ocean floor. They also study the boundary between the ocean floor and the continents, including continental shelves, estuaries, and bays.

Of course, landforms on Earth exist on land, but they can also be found under the water. Mountains, valleys, volcanoes, islands, plains, and canyons all exist in the ocean. In fact, Earth's highest peaks, deepest valleys, and flattest vast plains are all in the ocean.

Marine geologists locate underwater volcanoes by using sonar that bounces sound waves off rock formations. They also use remote sensing technology to map the ridges and valleys. The ocean bottom is one of the most active places on Earth from a geological point of view.

Mineralogists

Mineralogists study mineral formations. They observe minerals that have human value, such as **ore minerals**. Ore minerals are the source of building materials like iron, aluminum, copper, and zinc. Mineralogists use many tools, including picks and microscopes.

Atmospheric Scientists

Atmospheric scientists study the composition and activities of Earth's air. They may study weather and the effects of solar radiation on the atmosphere. They may also study atmospheric chemistry, including air pollution, global climate dynamics, and climate change. Atmospheric scientists use weather balloons, air samplers, and computer programs to collect and analyze data.

Seismologists

Seismologists are the specialists who study earthquakes. They use seismographs and computer programs to study fault lines. They place instruments along known faults to collect data. These data help them predict when large, destructive earthquakes might happen.

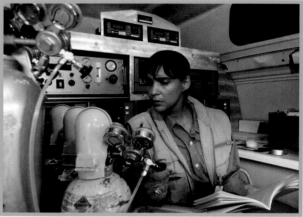

Structural Geologists

Structural geologists study the reshaping of Earth's surface. They also look at the massive forces that produce earthquakes and create mountains. They use maps and computer programs to study any changes to Earth's surface.

Hydrologists

Hydrologists inventory and monitor Earth's fresh water. They test water quality for pollution. They also study flooding and soil erosion. Hydrologists use tools such as flow meters, depth meters, and pH meters.

Petroleum Geologists

Petroleum geologists explore Earth for deposits of oil and natural gas. The oil and natural gas can be extracted for use as energy sources. Petroleum geologists analyze sound waves using computer programs.

Volcanologists

Volcanologists study volcanoes and possible volcanic regions. They use tiltmeters and computers to help predict when and where volcanoes might erupt and what damage they might cause.

Soil Scientists

Soil scientists study the composition and quality of soils. They look for ways to keep farm soils stable and fertile for growing food crops. They engineer methods to prevent soil erosion.

A cotton field with grass planted in rows to prevent wind erosion of the soil

Corn and alfalfa crops are planted in alternating rows to protect them from soil erosion.

A soil scientist field mapping soils using GPS technology

Soil Science at Home

Many home gardeners improve their garden soil with compost. Composting is a way to produce humus. Just take vegetable waste from the kitchen, lawn clippings, leaves, and other plant material, and pile it in a bin. As bacteria and fungus decompose the organic material, it slowly changes into pure humus. The humus can be worked into soil to enrich it.

Kitchen waste and leaves are collected and broken down in a compost bin.

After several months, the compost has turned into humus and is added to garden soil.

A home gardener uses compost in garden beds to enrich the soil with humus.

Garden plants growing in soil with humus are large and healthy.

Sand, gravel, and pebbles are the aggregates used in concrete.

Making Concrete

Concrete is a rocklike construction material that is made by people. You have probably seen a lot of places where **concrete** is used. Many highways are made of concrete. Bridges and overpasses are often made of concrete. Dams and stadiums are made of concrete.

What is concrete? You might see cement trucks at construction sites. They should be called concrete trucks because the material they carry in the big, round container is actually concrete. Concrete is a mixture of cement and aggregates. Aggregates are pieces of rock of different sizes. Small aggregates include sand and gravel. Larger aggregates include pebbles of several sizes.

Portland cement is a fine gray powder made from limestone. Limestone is dug out of a quarry. Then it is heated to a high temperature in a furnace and ground into a fine powder.

When Portland cement is mixed with water, it makes a sticky, mudlike mixture. Over time, the cement mixture cures (hardens). The mixture changes into a solid, hard lump. Cured cement is as hard as rock.

In order to take full advantage of this property of cement, aggregates are added to the sticky mixture. The cement bonds to the pieces of rock and sand, cementing them together into one strong mass. The mixture of cement, water, and aggregates is a thick fluid that can be poured into forms. The big container on the back of the cement truck is always turning around and around. The motion keeps the concrete moving around so it won't harden inside the truck.

Cement trucks usually don't travel very far from the plant where they load up with ingredients. Many contractors require that the concrete be poured within 90 minutes after loading. If the concrete hardens in the truck, it might be necessary to use jackhammers to break up the concrete.

Foundations for buildings are made by pouring concrete into forms. When the concrete foundation is cured, the building is constructed on top of the foundation.

Cement trucks deliver concrete to construction sites.

Concrete is poured into forms to make foundations for buildings.

Concrete foundations are different in different regions. For example, concrete in North Carolina is different from concrete in Texas, Wisconsin, or Oregon. That's because the aggregates mixed with the cement are always from the local region. It is too expensive to transport sand, gravel, and pebbles over long distances to make concrete, so the aggregates are local. Where did the aggregates come from to make the concrete foundation used to build your school?

Concrete can be used as stepping stones in a garden.

This school has concrete steps and walls.

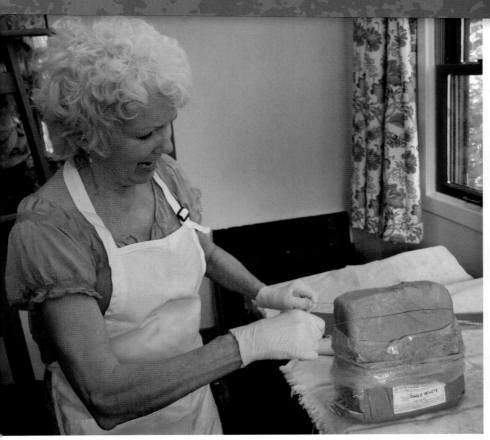

Cutting the clay with a wire

Rolling the clay flat with a rolling pin

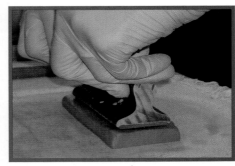

Using a stamp to press a shape into the clay

Earth Materials in Art

Rose Craig (1943–) is an artist. Rose makes beautiful watercolor paintings, and she is a skilled graphic artist and illustrator. For many years, Rose worked for the FOSS science program, drawing illustrations for articles in the *Science Resources* books. Rose has one other artistic interest, too. Rose makes ceramic tiles. Ceramic tiles are made of clay. As you know, clay is an earth material, so Rose is an earth material artist.

This is how she makes her beautiful tiles. First, she places a big pile of clay on her worktable. She uses a big rolling pin to flatten the clay into one big, thin sheet. The rolling process is very similar to rolling pie dough for a pie crust. When the clay is rolled out just right, Rose uses a straightedge and a knife to cut the slab of clay into rectangular pieces. Rose then uses rubber stamps to press a shape into the surface of the soft clay. She might use a fish shape, a flower shape, or a dragonfly shape, and then trim around the shape. Now Rose has to wait for the clay to dry.

A fish shape stamped in the clay

A kiln is a very hot oven that changes the clay into ceramic tiles.

Finished tiles

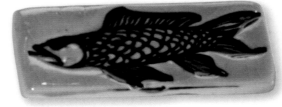

After 3 or 4 days, the tiles are dry and ready to be decorated with special paints called glazes. First, Rose paints a background color on the tiles. When that is dry, she presses the design into the clay again and enhances it with bright contrasting colors.

When the tiles are painted just the way Rose likes them, she puts them in a kiln. A kiln is an oven that gets really hot, much hotter than a pizza oven. The intense heat changes the clay into rock-hard ceramic tiles. The colored glaze becomes intensely shiny and hard as glass. The finished product is beautiful and useful. Because ceramic tiles are waterproof, they are good surfaces for sinks and counters that get wet. They are also useful outside in the garden or on a deck because sunshine, rain, or snow will not damage them.

Science Safety Rules

1. Listen carefully to your teacher's instructions. Follow all directions. Ask questions if you don't know what to do.

2. Tell your teacher if you have any allergies.

3. Never put any materials in your mouth. Do not taste anything unless your teacher tells you to do so.

4. Never smell any unknown material. If your teacher tells you to smell something, wave your hand over the material to bring the smell toward your nose.

5. Do not touch your face, mouth, ears, eyes, or nose while working with chemicals, plants, or animals.

6. Always protect your eyes. Wear safety goggles when necessary. Tell your teacher if you wear contact lenses.

7. Always wash your hands with soap and warm water after handling chemicals, plants, or animals.

8. Never mix any chemicals unless your teacher tells you to do so.

9. Report all spills, accidents, and injuries to your teacher.

10. Treat animals with respect, caution, and consideration.

11. Clean up your work space after each investigation.

12. Act responsibly during all science activities.

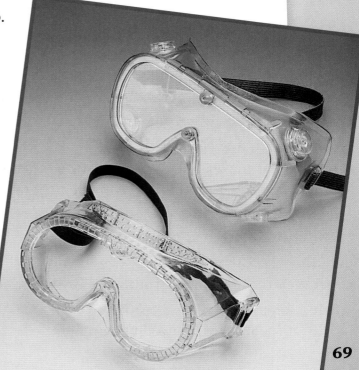

Glossary

abrasion the rubbing, grinding, and bumping of rocks that cause physical weathering

acid a substance that geologists use to identify rocks that contain calcite

bauxite an ore that has aluminum in it

birthstone a gem mineral that is identified with a month of the year

calcite a common rock-forming mineral in Earth's crust

chemical weathering the process by which the minerals in a rock can change due to chemicals in water and air. Chemical weathering can cause rocks to break apart.

cleavage the flat surfaces of freshly broken minerals

concrete a mixture of gravel, sand, cement, and water

core the center of Earth, made mostly of iron and nickel

crust Earth's outer layer of solid rock

decay when dead plants or animals break down into small pieces

deposition the settling of sediments

dissolve when a material mixes uniformly into another

earth material any nonliving natural resource that makes up Earth, including soil and water

earthquake a sudden movement of Earth's crust along a fault

erosion the carrying away of weathered earth materials by water, wind, or ice

fault a break in Earth's crust along which blocks of rock move past each other

feldspar a common rock-forming mineral in Earth's crust

flood a large amount of water flowing over land that is usually dry

fossil any remains, trace, or imprint of animal or plant life preserved in Earth's crust

fracture the uneven, rounded, or splintered surfaces of some minerals when they break

gem a hard mineral that can be cut into beautiful shapes

geologist a scientist who studies Earth, its materials, and its history

geoscientist a scientist who studies the use, distribution, and conservation of Earth's natural resources

glacier a large mass of ice moving slowly over land

gold a valuable dense metal that is found in ore and nuggets

granite an igneous rock that forms inside Earth

hardness a property of minerals determined by resistance to scratching

humus (HEW-mus) bits of dead plant and animal parts in the soil

igneous rock a rock that forms when melted rock (magma) hardens

landform a feature of the land, such as a mountain, canyon, or beach

landslide the sudden movement of earth materials down a slope

lava melted rock erupting onto Earth's surface, usually from a volcano

limestone a sedimentary rock made mostly of calcite

luster a description of the way light reflects off the surface of a mineral

magma melted rock below Earth's surface

magnetic a property of minerals that are attracted to magnets

mantle the solid rock material between Earth's core and crust

marble a metamorphic rock formed when limestone is subjected to heat and pressure

metallic describing the luster of a mineral that shines like metal

metamorphic rock a rock that forms when rocks and minerals are subjected to heat and pressure

mineral an ingredient of a rock

natural resource living or nonliving materials, such as soil, forests, or water, that come from the natural environment

nonmetallic describing the luster of a mineral that does not shine like metal

nutrient something that living things need to grow and stay healthy

opaque describing matter that does not let light shine through it

ore a rock or mineral that has a valuable substance in it, such as gold

ore mineral a mineral from which a valuable material, usually a metal, is extracted

particle a very small piece or part

physical weathering the process by which rocks are broken down by breaking and banging

property something that you can observe about an object or a material. Size, color, shape, texture, and smell are properties.

pumice a type of rock that forms when lava erupts from volcanoes

quartz a common mineral in igneous rocks

react to act or change in response to something

retain to hold or continue to hold

rock a solid earth material made of two or more minerals

rock cycle the processes by which rocks change into different kinds of rocks

sandstone a sedimentary rock made of sand particles stuck together

sediment pieces of weathered rock such as sand, deposited by wind, water, and ice

sedimentary rock a rock that forms when layers of sediments get stuck together

silt rocks that are smaller than sand, but bigger than clay

soil a mix of humus, sand, silt, clay, gravel, and/or pebbles

streak the mark left when a mineral sample is rubbed on a tile

transport to move or carry from one place to another

volcano an opening in Earth's crust where lava, cinders, ash, and gases come to the surface

weathering the process by which larger rocks crack and break apart over time to form smaller rocks

Index